Drones:

Mastering Flight Techniques

Copyright © Brian Halliday 2017

All Rights Reserved.

Table of Contents

Abbreviations vii

Introduction 1

The Basics 4

    Airspace Classification 4

    The Principles of Flight 6

    Thrust 6

    Drag 6

    Lift 6

    Weight 7

    Pitch Roll and Yaw 7

    Pitch 8

    Roll 8

    Yaw 8

    Weather conditions 9

    Wind and Rain 9

    Using Clouds to Predict the Weather 13

    Radio and Electromagnetic interference 16

Calibration and settings 18

    Ground Station (Remote Control) 18

    GPS Mode 20

    Attitude Mode 20

    Rate/Acro Mode 23

    RTH or Return to Home Mode 23

    Pre and post take-off checks 24

    Pre-flight checks and actions 24

Your operating area 28

    Situational awareness 28

    Take-off site and boundaries 28

    What's applicable? 29

    Drones and the Data Protection Act 31

Beginner's flight techniques 32

    Good Airmanship 32

    Controlled, Straights, Corners, Circles and Figure of 8 33

    Altitude 34

    Manoeuvres 34

    Drone Racing Manoeuvres 35

Filming and Photography 39

    Framing shots 40

    Frame rates 41

    White Balance (WB) 42

    Reveals 43

    General aerial filming techniques 45

    Moving Targets 47

    Mapping 49

    3D Models 53

    Structural survey 56

    Conservation 62

Drone Functionality 64

    Airworthiness 64

    What to Look/Listen out for 65

Drone Racing 66

Prepare for impact 69

Further Tips and FAQs 72

Afterword 76

v

Abbreviations

ATCU	Air Traffic Control Unit
ATTI	Attitude mode
ATZ	Aerodrome Air Traffic Control Zone
BNUC-s	Basic National UAS Certificate
CAA	Civil Aviation Authority
DEM	Digital Elevation Model
EASA	European Aviation Safety Agency
EMI	Electromagnetic Interference
EVLOS	Extended Visual Line of Sight
FOD	Foreign Object Debris
FPV	First Person View
GPS	Global Positioning System
MATZ	Military Air Traffic Control Zone
MET Data	Meteorological Data
NOTAM	Notice to Airmen
OSC	Operating Safety Case
OPS Manual	Operations Manual
PFAW	Permissions For Aerial Work
PFCO	Permission For Commercial Operations
RPAS	Remotely Piloted Aircraft System
RPAS-c	Remotely Piloted Aircraft System Chief
RPQ	Remote Pilot Qualification
RFI	Radio Frequency Interference
RTH	Return to Home
UAS	Unmanned Aerial System
UAV	Unmanned Aerial Vehicle
VLOS	Visual Line of Sight
WB	White Balance

Introduction

Thank you and congratulations on purchasing the second book in the series *Drones*, this book was created as a standalone book but also to work alongside and as a natural follow on to the first book in the series *Drones: The Professional Drone Pilot's Manual* and *Drones: The Professional UAV Pilot's Flight Time and Maintenance Logbook*. Together they will help you build and maintain a superior level of drone piloting skills, ensuring that you get the most out of your chosen drone whilst having as much fun as possible.

I'm going to be very clear here, READ YOUR INSTRUCTIONS MANUAL BEFORE ATTEMPTING TO FLY YOUR DRONE. This may seem obvious but we've all seen the Christmas day videos showing fresh out the box drones being obliterated on the concrete, as well as the downward spiral to earth caused by not properly fitting your props (propellers). No one wants to be that guy or girl posting that video. Pay your UAV/drone the respect it so rightly deserves and it will pay you back every time.

Drones sometimes get a raw deal in the media (in my humble opinion) due to the frequent proximity to aeroplane reports. Personally I follow up and research the reports of near collisions with planes, and trust me they are extremely rare (Unicorn rare). Almost every case that I have investigated turned out to be sensationalism, I assume to sell papers and generate clicks. In most cases it turned out to be a case of misinformation or mistaken identity and I am yet to find an official incident involving proximity to an aeroplane in flight. Unfortunately, the media tend not to follow up these stories. Pilots also, I believe at times face prejudices due to a misguided media who paint drone pilots as reckless individuals. My personal standpoint on this is that we all spend so much time and money on our aircraft that this alone should show our sceptics that our drones are viewed from a place of respect not disregard. Therefore, we both value and look after them. Individuals who build their own racing drones spend countless man hours on developing their aircrafts. Technically minded individuals like these do not endanger their own work. Unless they're in second place in the final bend!

That being said, it is only a matter of time before someone does do something stupid and the laws surrounding UAV and drone usage tighten up. As drone pilots, it is all of our responsibilities to act professionally at all times, and always operate our aircraft in a manner that is respectful to the surrounding community. By demonstrating that we are safety conscious, competent drone operators we will be in fact helping drones and the commercial UAV/drone services industry to evolve into something that everyone can believe in.

In this book, we will first go over the basics to ensure that you have a solid foundation upon which you can build your drone piloting skillset. I will describe the different flying modes along with external factors such as weather conditions and their possible impacts on your drone. You will learn a multitude of flight techniques that are effective in various settings as well as tried and tested methods to help you capture stunning aerial footage and accurate photographic data every time.

There's a chapter dedicated to drone functionality where I have included common problems and how to fix them, and of course *Drones: Mastering Flight Techniques* wouldn't be complete without spending a little time on Drone Racing.

I have written *Drones: Mastering Flight Techniques* in the hopes that it will inspire prospective and experienced pilots alike to further develop a full and effective UAV/drone piloting skillset. Commercial UAV Operators can look to this book for tips and ideas on flightpaths and positioning in order to achieve that perfect shot. There's no teacher like personal experience and the information within this book is sure to attribute to and enhance your experiences as a drone pilot whether you're a hobbyist or a seasoned professional.

If you enjoy this book or have any comments, please leave a short Amazon review. It would be greatly appreciated.

The Basics

There are many factors a UAV operator needs to consider before taking to the air. Here I will briefly discuss only the basic variables that directly affect how and when you should fly. Environmental aspects such as temperature and wind speed cannot be overlooked. Even a light wind can cause drone motors to strain and potentially overheat let alone the damage a sudden gust can cause. Within this section you will learn how to forecast the coming weather by reading clouds, followed by a description of and tips on how to use the *'Golden Hour'*. I won't dwell on the rules and principles of flight too much, but there are some fundamentals that you need to understand in order to become a master drone pilot. Underestimating the effects of these forces will seriously limit your potential as a drone pilot (costing you financially by way of replacement drones and endangering those with proximity to your operating area).

Airspace Classification

All of the airspace within the UK falls into 1 of 7 classes. The 7 classes of British airspace allow for the safe travel of thousands of commercial passenger flights each day. It is not just commercial airlines that operate within these classifications of airspace, aerodromes and the military also operate and control their own airspaces. All drone pilots must be aware of the classes of airspace in which they are safe to operate.

For drone users in the UK the only area in which we can legally fly (without further permissions from the CAA) is class G. Class G airspace is uncontrolled airspace, meaning that any aircraft may enter it and there are no restrictions on aircraft payload or flightpaths.

The classes of airspace A, B, C, D, and E are classed as controlled airspace. This means that they are under the control of ATC (Air Traffic Control) who, by analysing data from hundreds of flights and weather conditions, safely direct flights and decide the flightpaths taken above the UK. Further to the ATCU (Air Traffic Control Unit) there are nationwide ATZs (Aerodrome Air Traffic Control Zones) and MATZ s (Military Air Traffic Control Zones). When flying close to either an ATZ or MATZ it is a good idea to contact them prior to the flight to let them know you will be operating unmanned aircraft in the area, they too can let you know if they are conducting any operations that may affect or add to the limitations of your prosed flight. In my experience both ATZs and MATZs are hospitable towards drone operators as long as they are notified and kept in the loop.

To find out the classification of your local airspace download the Skydemon app or the free version Skydemon light. Another useful resource for information on airspace classification across the UK is notaminfo.com.

The Principles of Flight

The four basic effects that affect an aircraft whilst in flight are as follows: **Thrust**, the force that moves your aircraft forward, **Drag** that holds it back, **Lift** is the force that acts to keep your drone airborne and **weight** as gravity pulls it down.

Thrust

Thrust is described as the force that pushes an aircraft forward and this mechanical force is generated by either engines or motors. Thrust is required to overcome the *drag* force on the aircraft and compensate for the aircrafts *weight*.

Drag

Drag or air resistance, simply put is the affect the air has on your aircraft as it moves through it. *Drag* acts as friction to slow down the velocity of an aircraft. Understanding the nature of *drag* will allow you to use it to your advantage when taking long corners and 90° turns.

Lift

Your drone's *lift* is provided by the propellers which in turn are powered by motors. UAV/Drone propellers are much like an aeroplanes wing and shaped in a similar way. The official name for the shape of a wing or propeller is an Aerofoil. An Aerofoil is any structure that has curved edges with a ratio that favours *lift* over *drag* during flight. When your propeller blades move through the air they act as an aerofoil, creating the aerodynamic force of *lift*. A fixed wing UAV aircraft uses aerofoils to generate lift much in the same way as an aeroplane; most fixed wing aircraft also make use of ailerons. Ailerons are hinged control surfaces, usually fitted along the edge of an aircraft's wing. During flight, well controlled ailerons affect the aircraft along the *roll* axis (more on *roll* below) usually working in opposition to enable the aircraft to bank (turn).

Weight

Weight is best defined as the effect of gravity on an aircraft.

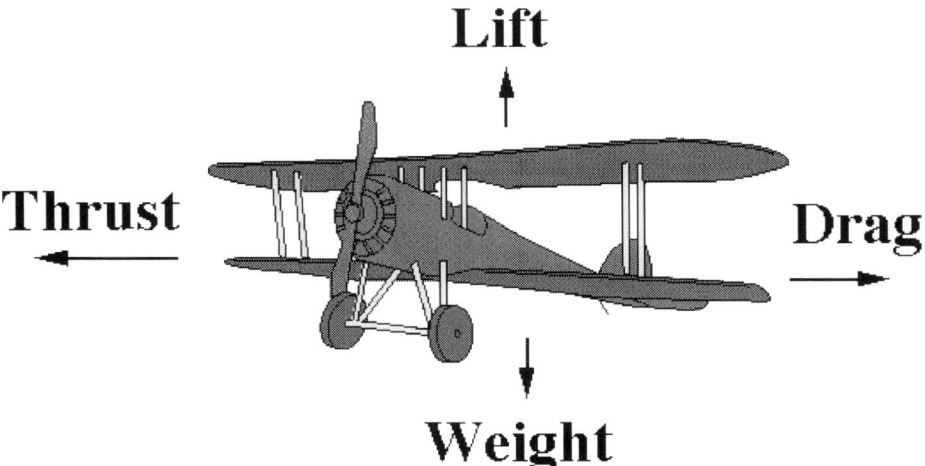

Pitch Roll and Yaw

Your UAV or drone operates along on the *Longitudinal, Lateral* and *Vertical axis*. If your drone is a quad or octocopter the *pitch, roll and yaw* will be controlled by your propellers and powered by individual motors. A fixed wing UAV will make use or elevators, ailerons and rudders to control the axis.

Pitch

Rotation around the *lateral axis* is called **Pitch**. Fixed wing aircraft use *Elevators* to affect pitch.

The angle (or **pitch**) at which a fixed wing aircraft's wings are tilted is called *the angle of attack*. Due to the *Coanda effect*, increasing the *angle of attack* creates greater downward air flow. This along with *Newton's Third Law of Motion (every action has an equal and opposite reaction)* generates more *lift*.

Roll

Rotation on the *Longituinal axis* is called **Roll** and is usually controlled by *Ailerons*. The *Roll* on the *Lateral axis* is the axis on which an aircraft turns/banks.

Yaw

When an aircraft rotates around the *Vertical axis*, this is referred to as **Yaw**. **Yaw** is generally controlled by a *Rudder system*.

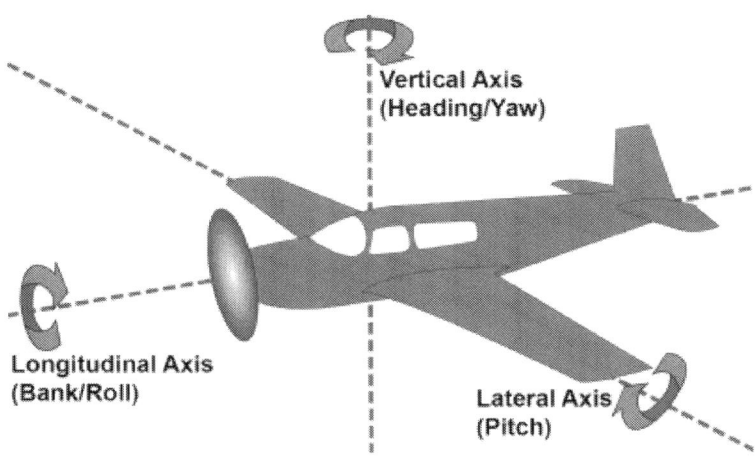

Weather conditions

There are a number of ways in which weather conditions can negatively impact your ability to successfully operate your drone. Aircraft operators are obligated by law to gather the relevant MET Data in order to conduct a safe flight.

Wind and Rain

With the potential to blow your drone in any direction without warning, wind is THE elemental force to be most concerned about. We can all tell if it's freezing outside or if it's raining, however judging how the wind is acting can be very tricky. Do not be fooled into thinking that the wind speed on the ground is the same as it is at 200 foot in the air. With altitude the wind direction can instantly change easily blowing a drone off course. Always use an *Anemometer* to gauge wind speed before a flight.

A headwind can cause your drone to gain or lose altitude as well as push it back. If this happens, especially if you happen to be using GPS assisted flight, the wind speed is too great and you should land your aircraft immediately. Racing drones are heavily affected by wind mainly due to their size and weight.

If ever in doubt, choose a take-off site that is facing the wind. Take off to an altitude of around 3 feet and watch the flight characteristics of your drone by flying forward, back, left and right to ascertain if it 'can handle' the current conditions. If your drone is being dragged off course, pulled up or pushed down you should land your drone immediately and "live to flight" another day.

Beyond the obvious reasons not to fly in the rain comes another reason many UAV operators are not aware of. The rain has the ability to absorb microwaves and all radio control drones use microwaves to communicate between ground station and aircraft.

Ice

Ice usually comes in 3 forms *frost*, *Clear ice* and *Rime ice*. All 3 can have a serious impact on your aircraft and I would strongly advise that you refrain from flying in icy conditions.

Frost- If allowed to settle on your drone's wings or propellers will cause extra drag and can also affect your air intake and rotor systems.

Clear Ice- The most dangerous type of ice, *clear ice* can cause serious airframe icing issues, leading to damage and the deformation of the aerofoils and/or ailerons.

Rime Ice-Is formed by the fast freezing of fog on a surface, like frost settling *rime ice* will affect the functionality of propellers and rotor systems. What makes rime ice particularly tricky is that it can form instantly, as water droplets, when cooled can freeze on impact.

Fog- Fog is formed in 1 of 2 ways, either by warm air passing over a cool surface or when warm air comes to rest and is cooled by a cold surface below, causing the water to condense. Once saturated the air becomes *fog*. It should be obvious but don't fly in *fog*, the water saturated air can enter your motors and air intake system causing critical damage.

Light

If you're using your drone for filming and photography, without doubt you will find yourself chasing the ever elusive Golden Hour. The golden hour is the first hour from sunrise and the last hour as the sun begins to set when the light is soft and diffuse, ideal for both ground based and aerial photography. When the sun is lower in the sky it produces fewer harsh shadows, allowing for a greater dynamic range without losing detail due to "white out". The golden hour is the perfect time for capturing stunning landscapes by drone; here are a few tips to help you along your way:

- Get there early.

- Plan your shoot, the light changes every ten minutes or so during the golden hour.

- Keep an eye on your camera exposure, for the above mentioned reason.

- Do not use your camera's "auto" white balance setting, experiment. Cloudy usually works best for me.

- Normally we avoid any lens flare but when used to full effect, golden hour lens flare can provide spectacular results.

- Try using a slow shutter speed when filming a moving target such as water, the motion blur effect that will be achievable adds a real sense of drama to a photograph.

- There is no better time for capturing the reflection of the sky in a body of water.

- Use foreground elements like trees to create a layered effect with serious depth.

- Use the shadows; the soft long shadows of the golden hour can look magical at an altitude.

- Consider not using an ND filter.

There's a cool app called Suncalc which provides daily information on the direction and time of the sun rise and set.

Below: Golden hour, Leicestershire.

Using Clouds to Predict the Weather

The presents of clouds signals a change in the weather and using or reading clouds is one of the most reliable ways of predicting the weather. There are up to 10 differing cloud types you should learn to recognise in order to be able to successfully forecast the coming weather. The 5 main identifying features of clouds are:

- Altitude
- Cloud Density
- Percentage of Cloud Cover
- Colour
- Shape

Below, I will go into more detail on some of the different cloud types. But as a rule to remember; the higher the clouds are, the better the coming weather will be.

Cumulus Clouds- Very easy to recognise, white, large and fluffy. Generally *cumulus clouds* are an indication of good weather, but many together may forecast coming showers.

Cirrus Clouds- *Cirrus clouds* form at a very high altitude and often appear broken and wispy. Cirrus clouds generally indicate fine weather ahead.

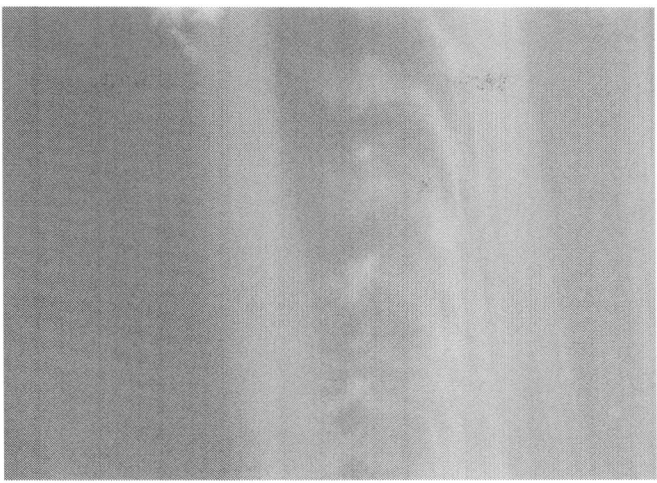

Stratus Clouds-These low clouds blanketing the sky often mean rain or a fog-ish drizzle. If these clouds have formed overnight, generally they will burn away as the sun rises.

Cirrocumulus Clouds- Are a sign of good weather and can often appear as high altitude ripples on a watery surface in the blue sky.

Altocumulus Clouds- Look similar to but much smaller than *cumulus clouds* and usually appear after a storm, forecasting fair weather ahead.

Cumulus Nimbus- These low altitude cloud giants often bring heavy rain, hail and thunder and lightning, they can be identified by their anvil shaped top.

Nimbostratus Clouds- Low lying clouds that tend to cover a large percentage of the sky, usually indicating heavy rain or snow is on the way.

Cirrostratus Clouds- Similar to the *cirrus clouds* but darker in colour; these clouds are generally made from ice particles and forecast rain or snow depending on the current temperature.

Altostratus Clouds- Resembling a grey fog in the sky and covering a large portion of the sky, *altostratus clouds* are a sure sign that rain is on the way.

Radio and Electromagnetic interference

Radio and Electromagnetic Interference is usually referred to as RFI (Radio Frequency Interference) or EMI (Electromagnetic Interference). RFI affects the way electronic circuits work and can seriously impact on the functionality of your drone. Even a car can affect your drone's GPS and compass calibration so always be sure to choose a suitable take-off site well away from any metal objects or objects that produce a radio signal. Consider purchasing a Spectrum Analyser to keep RFI risks as low as reasonably possible.

Serious RFI and EMI is often the cause for 'Rouge' or 'Flyaway' drone incidents. To limit these instances you should always calibrate your drones compass at every take-off location.

Any electrical device can cause radio or electromagnetic interference (including mobile phones). The main things to look out for are:

- Are all keys and metallic item well away from the drone?
- Do not take off from car roofs.
- Nearby gates and fences.
- Radio/Cell Towers.
- Power lines and Pylons.
- Underground Pipes.

If you are receiving electromagnetic interference alerts the first thing you should do is land and attempt to calibrate your drone's GPS and compass.

Try moving to a different take-off site and re-calibrating, if the alert persists I suggest contacting your drone manufacturer's technical support.

Calibration and settings

Your set up and calibration process will tend to vary from drone to drone depending on make and model, however regardless of your chosen aircraft, proper calibration and set up is essential. In most cases calibration involves linking your drone to the GPS network in order to gain GPS assisted flight. Further to GPS calibration IMU checks should be completed along with gimbal calibration to ensure a level horizon is observed. If your drone has its own "return to home" function, it's a good idea to set a return home altitude of at least 100 feet but again you must find out what works best for you.

There are a number of different flight modes for most drones, each with their own individual pros and cons. Below I will go into more detail on the varying flight modes available to most UAVs.

Ground Station (Remote Control)

A brief nod to ground station/remote control settings. Generally most quad and octocopters have 3 basic modes of control for the sticks (further modes are customisable);

Mode 1 Left Stick- Forward + Backwards- Yaw Left + Right

Mode 1 Right Stick- Up + Down (altitude) - Left + Right Direction (roll and throttle)

Mode 2 Left Stick- Up + Down (altitude) - Yaw Left + Right

Mode 2 Right Stick- Forwards + Backwards- Left + Right Direction (roll and throttle)

Mode 3 Left Stick- Forward + Back- Left + Right Direction (roll and pitch throttle)

Mode 3 Right Stick- Up + Down (altitude) - Yaw Left + Right

Some remote controls such as racing drone frequency transmitters have trim buttons alongside each control stick. Sometimes you will take-off and your drone will move any which way except straight up, this happens because the controls are unbalanced. If your drone starts to drift in any direction use the corresponding trim button.

Racing Drone Remote Controls

Racing drone frequency transmitters and remote controls can and do vary greatly and you want to buy a remote with plenty of programmable buttons, this will allow you a greater range for tricks. A quality remote control is one of the first things you are going to need when building your racing drone. For racing choose a remote control with as many channels as possible. Standard remotes may only have 4 channels which adjust pitch, roll, yaw and the aircraft's throttle. Channels beyond 4 will control further flight modes allowing for tricks.

Once delved into an entire book could be written on the different modes and set up techniques of racing drone remotes and their components. **Drones: Building Your Own Drone** is soon to be released and will cover, in step by step fashion the process of designing and building your own FPV racing quadcopter.

GPS Mode

I will not go into too much depth on how GPS assisted flight works. For UAV and drone pilots GPS assisted flight completely changed the game, allowing for a fixed position in a 3D environment fully controlled by the on-board processors. A correctly piloted drone is now like having a 400 foot jib on set.

However, GPS assisted flight is at times unreliable and should not be relied upon. Keep a keen eye on your drone if it begins to drift at all. If at any point you are in doubt of the GPS system's functionality you should switch your aircraft to ATTI (Attitude) or Racing Mode.

Attitude Mode

It is safe to say that *Attitude/ATTI Mode (or sometimes known as stabilised flight)* takes considerably more skill to master than GPS Mode.

Mastering *attitude mode* should be at the top of any drone pilot's to do list. *Attitude mode* makes use of *Gyros* and *Accelerometers* to provide stabilisation; however without the support of GPS your drone will drift considerably. *Attitude mode* is perfect for controlled drifting for long shots but you must pay attention to your aircrafts momentum.

Slow down when going into corners and stay focussed as you may need to manually compensate with your remote to control the drift. I would suggest that at least half of your flight training hours should be completed in *ATTI Mode*.

Pre-Programmed Flight

Pre-Programmed Flight by waypoints is an amazing UAV application. My first experience with this was working with the Ascending Technologies Falcon 8. This highly advanced UAV comes with a program that allows you to create a flightpath at home to later upload into the Falcon 8 UAV on-site. Since then I have tried and tested a number of mapping UAVs and software. The generic process of programming a flightpath for mapping is as follows:

Download an online map of the area to be mapped (usually from Map Box).

Set out a flight area, this is generally done by placing marks (4 minimum) on the map around your chosen operating area.

Create a flight matrix. There will be a button for either generate flightpath or generate flight matrix. This will overlay a zig zag pattern on the map which is the proposed flightpath. This flightpath can be changed and manipulated within the mapping programme's parameters. Choose a start point close to your take-off site and note the amount of batteries the entire flightpath will take to complete.

I suggest creating at least 3 flight paths per mapping site. The first flightpath will have a North to South orientation, the second with an East to West orientation and the third, a South West to North East orientation. This will allow for images to be collect from various viewpoints which will generate a larger number of keypoints when processed, giving greater depth and detail to the finished map.

After setting your flightpath and UAV orientation you must choose the optimal altitude and flying speed for the proposed task.

Now your flightpath, speed, altitude and orientation are set, you must now choose your photo overlap percentage (I recommend at least 85%) and optimum camera angle.

Your camera angle will be dictated by project, for mapping you will generally use a face down (90° angle) camera position. For 3D mapping and modelling position your camera at around 60°-70° angle and use a greater percentage of overlap for you images.

Check everything, check again and then save the newly created flightpath.

When on-site connect your laptop or tablet to your drone (once it has been properly calibrated) and upload your chosen flightpath to the drone.

After completing all necessary functionality and safety checks you are now ready to start your pre-programmed flight.

Once your flight is complete upload all of your images to your chosen processing programme, which will check for continuity between your images and alert you to any issues. All that is left is to export the files and deliver them to your client.

Rate/Acro Mode

Rate or *Acro Modes* are modes usually only used by racing drones. These unassisted flight modes disable any on board stabilisation software, meaning the pilot has the ability to aggressively turn and manoeuver the drone in order to perform tricks such as flips and high speed slaloms. *Rate Mode* takes a lot of practice to master and takes a completely new level skill compared to both *GPS* assisted and *ATTI Mode* flying.

RTH or Return to Home Mode

Return to Home Mode (RTH) is a very welcome addition to any drone. More and more drones are being released with this feature which when all else fails, you guessed it, returns your drone to home (the recorded position of the ground station or remote). In order to effectively use a drone's RTH function you MUST ensure your drone is properly calibrated. We've all seen the stories online of uncalibrated drones trying to "return home" to China when a pilot loses the data control link or presses the RTH button when they have lost sight of their aircraft (drones when flown should always be kept in line of sight). Prior to testing out the RTH feature it is important that you set a return to home altitude that is high enough to go over any expected hazards such as trees.

After initial testing to ensure the process of the RTH mode is understood and functioning correctly, it should not be relied upon; it should be used when needed as regular use of the RTH function demonstrates a low level of piloting skill. In the case of a complete loss of the data control link (communication breakdown between the remote control and aircraft) for any reason, the RTH function should automatically initiate and the drone will bring itself home.

Pre and post take-off checks

Pre-flight checks and actions

Checks

- Visually check the operating area/flying site for previously unplanned for/ unexpected hazards.
- Wind speed/direction/weather safe to fly?
- Other airspace/ground users.
- Local ATC frequency.
- Spectrum Analysis/safe to fly?
- Brief anyone on site of any potentially required safety procedures.

Actions

- Clear an even take-off site away from hazards, potential radio and magnetic interferences.

- Place take-off/landing mat on an even surface, oriented to North, South, East and West.

- Ensure your drone is a safe distance from your ground station/remote control.

- Connect your platform battery.

- Connect propellers (if applicable).

- Connect payload i.e. camera.

- Turn on controller, is it functioning properly?

- Turn on drone and connect to remote control.

- Is the ground station/remote control and drone connection optimal?

- Camera or payload functionality checks.

- Is your battery full?

- Calibrate drone.

- Take note the number of GPS satellites (the number of satellites required for a safe flight varies by drone make and model, 10 is usually a good number).

- "Safe to fly" checks on any apps or screens.

- Conduct a final visual check of your operating area.

- Pilot calls "Clear" before taking off.

Please note: The above are examples to be considered, not THE definitive Pre-Flight Checklist as each pilot must author their own checklists specific to their drone and the services they provide.

Post take-off checks

- Check UAV functionality by hovering at 5 metres, followed by moving forward, back, left and right to ensure stability in all directions before moving away from the take-off site.

- Any data downlink, monitor or FPV goggles are functioning correctly.

- Battery level check.

Post flight checks and actions

- Remove battery and check for visual damage and any signs of overheating.

- Remove propellers and check for any damage/wear and tear to propellers and connection points.

- Remove payload.

- Check rotor systems for any sings of damage, overheating or FOD (Foreign Object Debris).

For further information on Pre-Flight Checklists and FRCs (Flight Reference Cards) please see the first book in this series *Drones: The Professional Drone Pilot's Manual.*

Your operating area

Situational awareness

Situational awareness extends far beyond knowing the movements and positions of both pilot and drone, it is fundamental to the concept of good airmanship. Situational awareness is a skill that should be carried into everything you do. Being aware of any operating and situational limitations is important but even more so is knowing your own and your drone's limitations which are the true limitations to any situation.

Always conduct a brief walk through and mental risk assessment of your operating area prior to any flight.

Always choose your ground station site somewhere with the greatest field of view possible, preferable with a hedgerow or wall behind you so you know what is at your back at all times.

When flying your drone pay attention to any public footpaths and any areas likely that unexpected hazards (such as dog walkers) may appear from.

Take-off site and boundaries

- Your take off site.

- A 30meter secured take-off perimeter.

- A minimum 50meter secured outer perimeter.

- 2 Emergency landing sites.

- Ground station.

- Flight Safety Officer/Observers' position.

Take-off site and flight limitations

- Unable to fly 5km near airports.

- Flight limited to below 400ft vertically.

- Flight limited to 500m horizontally.

- Unable to fly within 150m of any congested area.

- Unable to fly within 50metres of any person, vessel, vehicle, or structure not in control of the pilot in charge.

What's applicable?

When flying a drone anywhere in the UK there are a number of laws to be observed beyond the obvious do not endanger aircraft, persons or property. Within the UK there are 3 classifications that any drone or fixed wing UAV will fall in to:

Class 1 Small Unmanned Aircraft- To fall into this category the UAV must weigh below 20kg.

Class B Light UAS- These unmanned aircraft are between 20Kg and 150kg.

Class C UAS- Aircraft over 150kg.

The vast majority of drone pilots whether racer, commercial or hobbyist will be flying class 1 category drones which must be kept within the pilot's visual line of sight or VLOS, unless permission has been granted to extend operations to EVLOS (Extended Visual Line of Sight) by the CAA.

It takes a combination of skill and focus to pilot a drone, when flying always bear in mind:

People <u>WILL</u> come over to speak to you about your drone and ask the usual questions, how high? How far? Make sure you know the answers to these questions, explain that you have permission to fly in this area and politely let them know that piloting the drone takes up all of your concentration.

Always take-off from a flat stable surface away from any metallic objects.

If the drone's orientation and heading are facing you this will mean the drone will move in the opposite direction to the movement of the directional control sticks.

Pay attention to your flight time as well as the drone's distance from the ground station, nobody wants to run out of power 200 meters from "home".

At altitude wind speed and direction can change in an instant.

When possible measure the winds speed prior to flight, any wind speeds of over 18mph are above the limits of most drones.

Battery output levels.

Cold temperatures greatly affect battery performance.

Be aware of any local wildlife and the reactions they may have to the drone whilst in flight.

Stable, controlled flight produces the best footage and stills.

Drones and the Data Protection Act

Drones with cameras attached can be classed as surveillance cameras and in such instances the Data Protection Act will apply, so download a copy of the act and comply with any guidelines laid out. Generally be aware that the act covers the filming of individuals without consent. A conscious drone pilot will also keep a safe distance from other people, properties, schools and local business. The CAA regulations mentioned above cover the physical distance limitations but remember that your camera's vision is far and wide and many people are uncomfortable, even if you aren't anywhere near their property.

If someone happens to pass into shot 50metres away it is unlikely that the individual could be identified on the footage, however with that being said you should always operate your drone in a way that is respectful to and well away from other ground users.

Beginner's flight techniques

Good Airmanship

Good airmanship is a state of mind not just a set of rules and checklists. In its essence it's the most important part of your skillset, it is your skillset. Starting from a place of good airmanship will ensure everything you do carries safety risks that are as low as reasonably possible. Regular scheduled maintenance is an integral part of good airmanship and developing a process to maintain your aircraft's airworthiness is essential, but more on this later. All drone pilots must operate safely and within the law but no qualification is required for the drone enthusiast, permission from land owners however must always be obtained and the flight limitations mentioned above must always be observed.

- If you are a Commercial UAV Operator or plan to operate drones commercially the criteria of good airmanship includes:

- Obtaining a recognised RPAS or BANUC-s Qualification by completing a ground school course followed by flight test.

- Author a complete set of FRCs (Flight Reference Cards) and UAS Operating Safety Case.

- Gain your PFCO (Permissions For Commercial Operation) or PFAW (Permissions For Aerial Work) from the CAA (Civil Aviation Authority).

- Keep Valid Insurance.

- Keep and maintain a complete and up to date flight hours logbook, maintenance logbook and battery charge logbook.

- Always gain permission from the landowner before beginning any aerial work.

Controlled, Straights, Corners, Circles and Figure of 8

When taking off, you should only need to use your left stick. Slowly pushing the left control stick forward will engage your drone's motors causing the propellers to spin. As you push the left control stick forward get a feeling for the sensitivity of the throttle as your drone lifts off and starts to gain altitude. Rise to above 3 feet as the drone may be affected by its own wind disturbance at below 3 feet.

Begin the controlled straight by gently pushing forward on the right control stick (pay attention to any drop or gain in altitude and pitch). To turn, push the left stick to the left to yaw left and right to yaw right, as pressure is applied either left or right on the left stick the drone will turn/yaw to its new heading. Continue a controlled straight from here and repeat the above until you have completed a square. It will take practice to get a precise looking square but you will master this in no time.

To complete circles and figures of 8 you must (whilst continuing to push forward on the right stick), keep light pressure on the left stick, to the left or right depending on the direction of your circle/figure of 8. Constant pressure on the left stick is not required, as slight yaw every few seconds should suffice but this will be dictated by size of the circle and the speed at which the drone is flying.

Altitude

In addition to the x (horizontal), z (forward and back) axis, drones also operate along the y (vertical) axis. Controlling altitude efficiently takes time but the best way to practice this is to go back to practicing squares. After each corner of the square, as the drone goes back into a controlled straight, push forward on the left stick to gain altitude. Do not release the altitude throttle as you come into the corner, instead use the left stick to yaw into the corner as well as gain altitude at a controlled rate. Once a sufficient height is reached, continue the squares but gradually lose altitude as you complete each square.

Manoeuvres

The 3 point turn manoeuver is the same basic movement as is done in a car. Start by hovering with the drone. Lightly pull back the right control stick causing the drone to slowly move backwards. Whilst the drone flies backwards begin to push the left stick to the left or right causing the drone to yaw slightly then release. As the drone backs into the new heading release the right stick, then push forward on right stick and turn out of the 3 point turn with your left control stick.

To perform a reverse circle pull back on the right stick to move your aircraft backwards. As the drone moves, gently control the yaw on the left stick in order to fly a backwards circle. Once you can do this in a controlled fashion it's time to practice circles whilst keeping your camera facing a set location or direction. This is much harder than it sounds. Continue on to "rolling" circles where the drone in effect flies sideways and the pilot controls the turn by yawing with the left control stick. Learn to complete accurate circles this way and you are well on your way to *Mastering Flight Techniques*.

Drone Racing Manoeuvres

Racing drone manoeuvres are different to the manoeuvers mentioned above as racing drones are mostly flown in *rate/acro mode*, where flying unassisted and at great speeds performing seemingly simple manoeuvres takes great skill. Here we will cover a few basic moves to practice beyond the controlled hover. The remote controls can vary greatly so in this section I have kept the details down to the basic sticks, for further info on racing drone remotes please see the *calibration* section of this book. The manoeuvres detailed below are all from the prospective of a pilot flying with the control sticks set to *mode 2*.

90° and 180° turns sound simple but pulling either off at 60mph-80mph is difficult to say the least. Both the *90° and 180° turns* start with a controlled straight by pushing forward on the right stick. As you come to your turn slightly relax the pressure on the right stick and gently push left on the left stick which will cause the drone to yaw to the left, release the pressure on the left stick once the drone is at the desired heading 90°. At this point you are going to have to use your right stick to compensate for the drone's current trajectory. As the drone turns to face its new heading it will carry momentum from the previous straight. In order to stop the drone drifting off course you must, (whilst continuing to add light pressure to the right stick throttle) push the right slightly to the left for a second or 2 before relaxing back into the controlled straight. It takes a real master to pull this off at high speed so do not be discouraged if it's a bit of a struggle to get right.

For the *180° turn*, follow the above but yaw to 180°, you must relax the right stick's throttle as you turn, the momentum will be accounted for by reapplying the forward pressure to the right stick once the 180° heading has been met.

Once you are comfortable with the *180° turn*, instead of applying pressure to the right stick in order to go forwards, pull it back as to perform the *180° turn* into a reverse. This is a bit trickier to pull off but is a really cool move.

Slalom is a series of consecutive turns around a row of obstacles that are usually placed in a straight line. These lines can be very tight or loose. First of all lead into your slalom from a controlled straight, eventually what happens next will come down to your own particular preference but beginners should concentrate on keeping on course and maintaining tight corners and lines.

An *x axis roll* can only be complete when the drone is in rate or acro mode which allows for more aggressive movements. While flying forwards (by pushing forward on the right stick but lowering the pressure coming into the trick) aggressively push the right stick to either the left or right depending on which way you want to *roll*. You will only need to keep the sideways pressure on the right stick for a moment, as the drone completes the flip release then push forward on the right stick to pull the aircraft out of the *roll* and back into the straight.

When completing an *x axis roll* the drone will lose considerable altitude, this can be used to your advantage when dropping levels on a 3D course.

If you want to keep your current altitude throughout the *x axis roll* you must, just before pushing left or right on the right stick, push forward slightly on the left stick to gain a slight jump in altitude to compensate for the drop from the roll. As the drone rolls release pressure on the left stick, push forward on the right stick and pull out of the roll.

Flips are a bit harder to master as the pilot's view changes massively through the flip, especially as you lose view of the horizon. With the *x axis roll* the horizon is always in view so it's pretty easy to keep your bearings. As you flip forward or backwards along the z axis the horizon disappears and the ground will come fully into view, followed again by the horizon. At first this is very disorientating; completing a few hundred flips will allow you to become accustom with the dizzying effect of *flips*.

Completing *flips* is very much like the *x axis roll* except instead of pushing hard left or right on the right stick you instead move the stick either forward or back. It is important to release all pressure on the right stick just before you push forward or back to *flip*. A little push forward on the left stick will ensure you do not lose any altitude through the *flip*.

Filming and Photography

Learning Photography and Cinematography can take years as can the learning curve to becoming a master drone pilot. Combining the 2 to capture professional grade footage and stills is no small ask. Capturing brilliant drone footage requires a lot of focus and attention to detail. I imagine we are all familiar with the dreaded "white-out" ruining our transition shots as well as having the perfect shot, if only you had had your drone positioned 5 metres to the left or right. Properly operating a drone takes heaps of concentration and a shot that by a glance at the data link looks amazing may, when viewed post flight, be full of mistakes.

The best way to combat this is to internalise a basic set of rules that will help you to always get the best out of your drone footage. This isn't a photography class, just a few simple things to remember that will make a huge difference.

- In the shot, place your subject in open space in relation to the background. An advantage of drone photography is that we can go that little bit higher than a standard photographer could, this gives added options for using angles to cut out undesirable aspects of a background. 1 cool effect is to shoot from around 7 metres and angle the camera as to make the ground the background; this stops any distractions above the horizon getting into your shot.

- Location is (almost) everything, has everyone had their fun filming fields and bushes? The odd tree perhaps? How about shots at your local reservoir? We've all done it and it's time to move on. As drones become more widespread the creative drone operator must widen their horizon (no pun intended) and begin to plan entire videos. A lot of this will involve travelling to different locations across the country, continent or if you're lucky the world. In order to get the superior landscapes that are so often the jewel in a promo video, the creative drone pilot must venture beyond their back yard.

- Where possible shoot all photographs in RAW format. This will allow a greater range in the editing suit.

- If in a pinch, adjust your shutter speed to compensate for any white-out (white balance) problems you are having. This wouldn't be advised by the professional photographer; however when you are at an altitude of 350feet with a battery that lasts 15 minutes tops, upping or lowering the shutter speed to gain or lose light is a quick fix. An ISO of 100 is recommended along with this method.

- If the quality of the shot is paramount, try using a camera operator (if your drone allows this option) to control the camera, allowing the pilot to fully concentrate on a smooth stable flight.

Framing shots

Properly framing your shots and subject will undoubtedly allow you to capture more dynamic images and footage. For any beginner, your natural inclination is going to be to place your subject slap bang in the middle of the page, do not do this. Most apps have an option to overlay a grid on the monitor (app screen), splitting the screen into thirds, usually 9 squares. Try having your subject in line with where the third point lines cross, this leaves room to "look" around the shot.

Photographs can always be cropped in post flight editing in order to achieve the "thirds" effect.

Another framing technique is to look for any vertical or horizontal lines in the fore or background (make sure the lines are straight and or level). Look for lines that lead into or out of your shots, treelines, roads and fences are good "go to" examples.

Framing a shot is the perfect way to give it context. This technique of drawing attention to your focal point by blocking out the outer edges of the shot via a foreground obstruction can have a bunch of benefits such as:

- People will instantly be drawn in to this kind of image.
- Adding an extra level of depth to your image.

Frame rates

Be wary of drones and cameras offering 4K and beyond, it may be true that the camera does in fact shoot in one of the many 4K modes but in truth without a proper frame rate of at least 30 frames per second your footage will come out juddering and jumpy. 1080HP with a frame rate of at least 30fps (frames per second) will look buttery smooth when compared to 4K footage filmed in 27fps. Unless you have the equipment to view and edit 4K footage I would strongly suggest filming in 1080p or 720HD with a high frame rate. This will give you a better overall quality.

White Balance (WB)

White balance is the software in your camera (and editing programmes) that eliminates by removal an unrealistic colour tones that are too bright or cold. Most drones on board cameras have at least some of the following white balance settings:

- Auto WB and Custom

- Neon

- Incandescent

- Sunny or Daylight

- Cloudy

Auto white balance uses its best guess to balance effects in the range of colour temperatures. Custom allows the operator to set the balance instead of relying on auto white balance's guess algorithm. Whilst remaining white balance modes are all set to differing colour temperatures.

As a minimum remember, background, frame rate, white balance and shutter speed.

Reveals

For your *standard reveal*, take off with your camera at a 90° angle (facing straight down) and slowly rise. It is important that you do this as smoothly as possible without any jumps or knocks. As the drone gains altitude it will begin to reveal the surrounding area, allowing you to turn a shot that at first glance seems close up into a grand reveal with little effort. As you reach your required altitude try gently panning up with your camera to slowly bring the horizon into view, if you get this right you can capture some brilliant "golden hour" shots relatively easily.

This next reveal is slightly different to the standard reveal but in my opinion the *45° degree reveal* has a greater overall impact and has more creative applications. Again after take-off you remain relatively low, focussing on your subject. This time, keeping your camera on your subject, fly your drone up and away from the subject at approximately a 45° angle which will gradually reveal a classic land/cityscape. Play around with this shot, see what angles work best for you and practice what speed you need to fly at to make the most of your chosen angles. At the risk of repeating myself, a slow and stable flightpath will almost always provide the best results.

Flyover shots are a tried and tested client pleaser and really simple to get. Position your drone in front of your subject at your desired altitude. With your camera facing the horizon slowly fly a straight line towards and over your subject. As you approach the subject gently pan down with your camera from 45° to the 90° angle so your camera is facing straight down just as you pass over the subject. Continue to fly for a further 20 metre or so (if it is safe to) to ensure you have the full shot you need.

The method to capturing a quality *reverse fly over reveal* differs to the reveal methods mentioned above. For this shot your drone's start position should be where you want your shot to finish. Frame your subject (leave your camera in this position), and start to fly towards and over your subject. Once you have flown past your subject you have 2 options. 1 You can fly backwards along the same flightpath, this gives you 2 shots to pick from (remember you can reverse your footage in post flight editing). Or option 2, turn your drone around and frame up the shot again from the opposite side and take the same shot from a different viewpoint allowing you more overall footage to choose from.

Next page: For this shot I started low, following visitors into the building, followed by a gradual rise whilst continuing a slow straight. As the visitors entered the build I gained enough altitude for the fantastic building behind to come into view. The still looks impressive but the raw footage was the "wish list" shot for the client and it came out great.

General aerial filming techniques

The Push in method is similar to the above mentioned *45° reveal* but in reverse. You begin by positioning your drone at a reasonable height and distance from your subject, and then simply fly towards your subject until they're perfectly framed.

Another move that's reasonably easy to pull off is *the descent and pan*. From a height, slowly descend and match the speed of the drone's descent by panning up with the camera from 90° (straight down). The effect from this type of shot is stunning and works well as a transition between scenes. Also the reverse of this, the ascent and pan has great effect too.

Flying a straight line at speed can give a cinematic feel to a shot when a subject is filmed at a high frame rate (from a distance) and the footage is slowed down in post flight editing. This will cause your subject to move in slow motion whilst the camera still appears to be moving through the landscape.

The bird's eye spin or spiral effect can be created by positioning your drone directly above your subject with the camera facing straight down. Very slowly, push on the left stick to begin to gain altitude whilst at the same time very gently push left or right also on the left stick. In effect you will be pushing slightly up and slightly left/right on the same stick, this is tricky to get right but can be mastered in a short time.

Drifting at altitude with the camera facing down can also be used to create excellent fade out footage for videos. Drifting above a forest in autumn for example provides beautiful results as you view the trees from directly above. Following a moving target from directly above in this way, especially when combined with a parallel follow (mentioned below) is defiantly something everyone should try.

The side to side is accomplished by slowly flying sideways to gradually bring your subject into shot. Continue to fly the same trajectory and speed until the subject is no longer in shot. Or you could try adding camera movements as to keep the subject in shot as the drone moves further away from it.

The 360° altitude pan is simple yet very effective. At your desired altitude (usually the higher end of the CAA set ceiling limit) gently yaw your drone either left or right using the left control stick. Do this as slowly as possible with a camera angle of around 50°-60° to get the smoothest effect.

Flying close to the ground at high speed is not something you should try on your first day flying your new drone but once you are comfortable with your aircrafts capabilities, try this out. In the right landscape the footage can look superb.

The sit and spy is probably the easiest technique contained in this book. Hover your drone in your desired position and have your subject move through the shot whilst the drone and camera stay still. You could add to this method by moving the gimbal to allow the camera to follow the subject.

POV filming style is perfect for a tour style video along a pathway or through a tunnel. Just fly in a controlled fashion along the "path" with your camera facing straight forward, slowly taking any corners or bends at the same speed a walker or jogger would. These shots look great when speeded up.

Moving Targets

The pursuit style shot is ideal for filming cars and when performed properly with planning is a safe and trusted method for filming moving targets. First of all position your drone in front of the subject, then turn around so you are facing the opposite direction. Position your camera turned backwards, allowing you to frame your subject. Begin to fly forwards as your subject begins to move and remember to keep the subject properly framed at all times. Assistants/spotters should be used during this kind of shoot along with a camera operator to ensure the pilot can focus their full attention on the safe piloting of the drone. How close, how fast and how high you fly will be determined by the subject's speed and the surroundings.

The pursuit method also works with the drone chasing the subject, however these shots should always be planned first to make sure the pilot and ground station are properly placed for the best view of the drone at all time. For experienced, pilots proper positioning of the ground station (remote control) almost goes without saying but capturing moving (sometimes fast moving) subjects is inherently more dangerous than filming standard footage so I have mentioned it here.

The Parallel follow is an excellent shot, which can be used to great effect when done right. The correct way to parallel is to position your drone at a 90° angle to your subject and turn your camera to face (and then frame) your subject. This allows you to fly the drone forward as oppose to having your drone face your subject and then fly to the left or right, which can be tricky when trying to fly a straight line.

Orbiting 360° (sometimes known as 'point of interest') around your subject is an extremely useful flight technique which has many applications beyond videography. Performing a controlled circle around your subject allows yet another shot unique to drone operators. Flying 360° works equally well for both stationary and moving subjects.

Try combining techniques to add even greater value to your piloting skillset. Whilst following a moving target, transition into a 360° perhaps let your subject drift out of shot, or ascend to a horizon shot. The end results are pretty much money in the bank. The limit is your creativity and the more time you spend practicing, the sooner you will become not only a master drone pilot but also gain a wealth of creative and technical knowledge.

Mapping

Successfully collecting data for creating maps (sometimes called raster data) usually involves zig zagging across the sky taking photos at set intervals which you later process to form a composite map. When producing maps this way you are able to embed geospatial data such as longitude and latitude. This method can also be used to create DEMs (Digital Elevation Models).

Most mapping flights are conducted using Pre-Programmed Flight Apps that ensure continuity between all the photos taken during the flight. I have tried out a number of these apps when building maps, all of which allow the pilot to choose a set height and speed for pre-programmed flight along with the number of photos taken and overlap of the photos taken.

From my experience I would suggest flying at an altitude of between 100 feet and 130 feet with an overlap of between 85% and 95% (this can vary by project). The apps will advise you on the number of batteries that will be required to complete the programmed flight and the proper process of returning to home for a battery change before returning to where the pre-programmed flight path previously left off.

During 2016 I worked with wildlife rangers to map local nature reserves in order to aid with planning and the general maintenance of the reserves. Both examples (images below) are of the same 67 acre area, a country park and Environmental Centre in Leicestershire. I have included the flight and image information as a worked example:

UAV- DJI Inspire 1 with Z3 camera payload-Altitude 160 feet/50 metres-speed 7 metres per second-Image overlap 90%-Camera angle 90°-Batteries used=10

Number of images taken 994-Mapping app used=Maps Made Easy-Processed by Maps Made Easy online-Processing time 4 hours- What I got paid….yeah right.

Bellow: The map of the Country Park, Environmental Centre and Nature Reserve. On the far left corner you can see the newly planted orchard that is beside the environmental centre. All the paths are clearly visible even through the wooded area as the map data was collected in early March before tree cover could return.

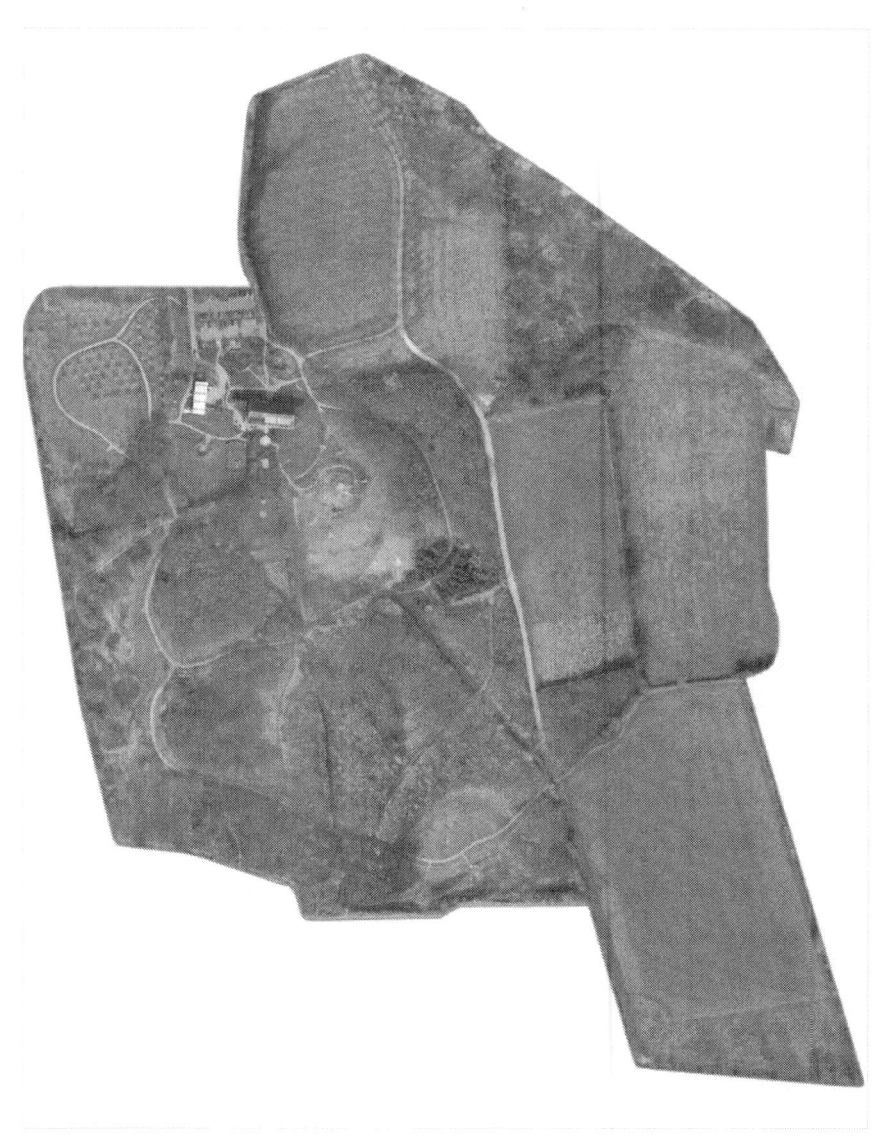

Below: A DEM (Digital Elevation Model) created with the same data as the above map. This DEM was created to help wildlife rangers in the planning of new water run off trentches to combat flooding. The original DEM data file can be zoomed in on many times over to be veiwed in extreme detail. The ripples in the first field show the remnants of tilling earth by oxen, which was last done on this site in the 1850s.

3D Models

Further To DEMs and 3D maps another service gaining commercial momentum is data collection for 3D imaging and modelling. This is a really cool area to get into, however accuracy is everything here and the key is to have as much overlap between your images as possible. Capturing the data for a 3D model is much the same as for mapping, although the processing methods can vary. Your chosen processing programme will sort through and find thousands of common points in your images, a number of these matches, the common points (called control or keypoints) are then correlated, then overlapped and a 3D image is generated.

Once you have captured and checked your data, you can then run it through a slicer and print your model off on a 3D printer.

Next page: A storyboard I put together demonstrating the process of 3D modelling by drone, the 3D images were taken by the Ascending Technologies Falcon 8 UAV and processed using Pix4D.

Image 1 is a true photograph of the life size mammoth.

Image 2 is the 2 man flight team. 1 UAV Operator and 1 Camera operator using FPV goggles.

Image 3 shows the angles from which the data (images) for the model was captured along the orbital flight path.

Images 4 and 5 show the 3D Model half way through processing and the fully finished 3D model now ready to be fed through a slicer programme and printed using a 3D printer or used for animation.

The examples on next page were created in order to view the structure from all angles prior to demolition. Unlike the above example the data for the images below were captured with a DJI Inspire 1 with a Zemuse3 camera. A standard "point of interest" or orbital flightpath was flown at 3, 5, 8, 10 and 12 metre altitudes and at varying distances. Post flight the 3D images were processed and created by Pix4D.

Below: The original

Below: The 3D model

Structural survey

The aerial surveying and inspection of a structure or construction site can be both expensive and time consuming, especially if data is required throughout the project. UAV operators are now able to provide a solution that reduces safety risks, costs and is time effective. Through utilising state of the art technology and high resolution cameras UAV operators now have the ability to produce extremely precise geo-referenced *Orthomosaics* (geometrically corrected aerial photography) and digital elevation models (DEMs). The processing programmes can formulate a number of post processing reports, allowing accurate and detailed information to be viewed at a glance. Footage can be viewed in real time via downlink to ensure you get the photographs you need every time. All data can be provided to the client in RAW format or fully edited by your in house team.

- Site Identification, set-up and analysis.
- Access to otherwise inaccessible areas.
- No need to close the site/structure.
- Safer-low risk to both workers and structures.
- Fewer EH&S issues.
- Lower costs.
- Excavation documentation.
- Site fly-through animation.

Drones can fulfil the need for a rapid though accurate documentation of objects, be it during an excavation, which is a dynamic process and therefore requires fast and preferably non-immersive documentation techniques which can also be suited to cover larger areas.

- Mapping, including Longitude/Latitude and measuring tools (Geospatial Data Collection).

- Proximity Analysis.

- 3D Photogrammetry (laser scanning alternative).

- DSMs (Digital Surface Models) including ground elevation.

- Provides an extra layer of detail beyond photography and can be combined with other layers to create hybrid maps much in the same way as google maps. Interactive 3D renderings and models can also be created.

There are a number of techniques that you need to practice to successfully complete a professional structural survey. You will be required to capture data "close up" in order to get the extremely detailed structural images that are expected. A professional structural survey takes more than your standard operator. It may take hundreds of hours in practice flights before a pilot is proficient in the controlled flying style needed for a competent survey.

Being a professional drone pilot alone will only get you so far in this field, to provide a proper survey you must know what you're looking for, so study up. No one is saying you need a degree in structural engineering, however demonstrating a basic level of industry knowledge and terms is a must.

Flying 360° around a fixed point is an extremely useful tool for surveys; it allows instant access from all sides to difficult to reach areas.

When conducting a general roof survey, start by flying a + pattern followed by an x pattern whilst positioning your camera at a 65°-75° angle. I would suggest flying at around 10 metres above the roof you are surveying, however this can vary by project.

Once you are satisfied you have captured sufficient footage and photos, you can begin to capture the structural integrity of the Gables (the upper section of the outer walls) and Flush eves (the lowest edges of the roof that usually protrude but can sometimes run flush to the roof itself).

Fly a straight line parallel to, but 3-5 foot lower than the roofs edge (to allow you a clear view of the guttering or any other fixed features). For this shot I would advise flying in the same way as *the parallel follow method* mentioned above.

As a worked example I have included some shots from a year-long project where I was tasked with the documentation of the progress of a new hospital throughout construction.

Below: An aerial view of the entire site.

Below: 3 Months into construction the building was really taking shape but the courtyards and outer gardens were yet to be started. This is a huge hospital that cost in excess of £50 million to construct and was designed to care for vulnerable young people.

Below: A bird's eye view of the entire building, the area was still a building site; at this point the inner courtyards, vegetable gardens, basketball/tennis courts and outer gardens and carpark were yet to be started. The client was so impressed with the shot below they request that we help with a further 2 videos, an information film about the healthcare group that built the hospital and a tour video of the new building upon completion.

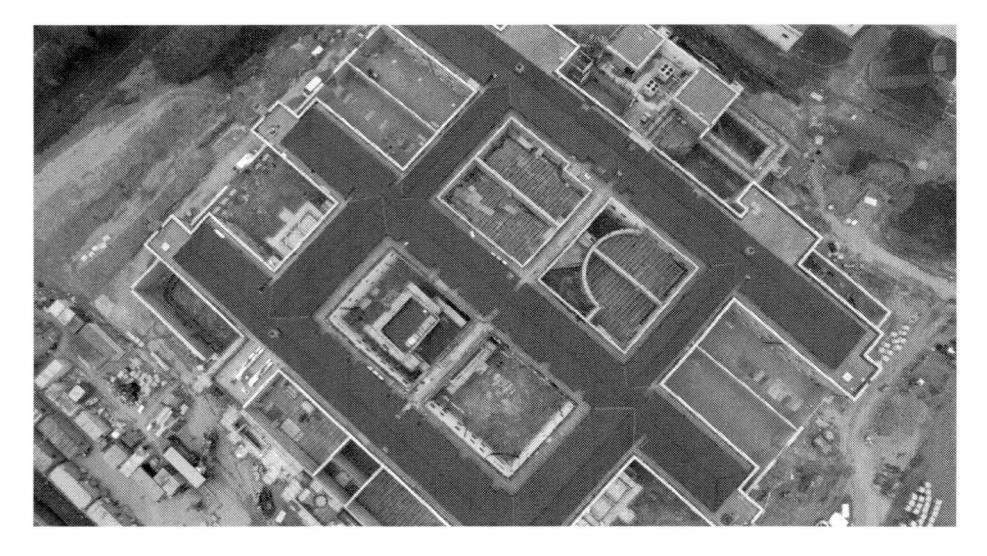

Next page: I have included few choice shots of the hospital so you can see how it looked upon completion of the project.

Below: Here you can see the colourfully designed courtyards, basketball/tennis courts and the outer gardens of the hospital.

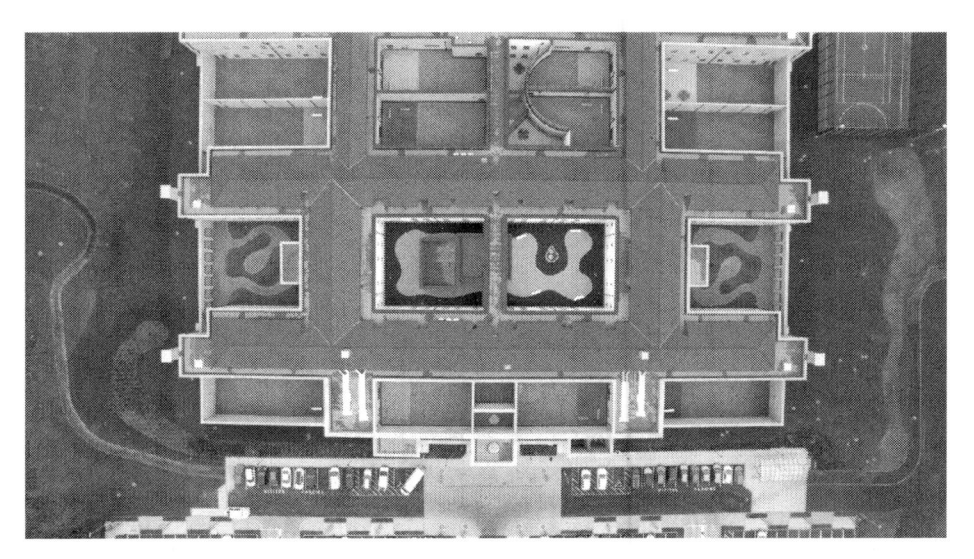

Conservation

We all aim to do our bit in the preservation and conservation of our natural resources as often as possible. UAVs are non-intrusive and able to discreetly monitor wildlife habitats without alarming the inhabitants or disrupting the surrounding areas and until recently, monitoring tree populations and density from the air would be a lengthy and expensive exercise. However, now with the aid of drone operators, conservation groups are able to successfully map huge areas in an afternoon. This low cost and environmentally friendly alternative to using a helicopter has the scope to provide an array of services in the name of conservation. Getting involved with conservation groups, even on a voluntary basis can lead to many opportunities.

Landowner's Permission

Whether it's a practice flight or a commercial project, obtaining the landowner's permission to operate (fly your drone) before beginning any task is essential. During the qualification stage of any project the landowner's contact details will need to be taken and permission granted before any project can be booked in. I would suggest attaining a signed *Landowner's Permissions Form* of some sort prior to flight of any kind.

Further Pre-notification (Commercial UAV/Drone Operators)

In some cases further pre-notification of certain organisations may be required. In all cases of a commercial drone project the relevant NOTAMs should be filed. All NOTAMs and notifications should be completed at the earliest opportunity (usually 28 days prior to project's start date).

For a more in depth look at NOTAMs please see the first book in the series *Drones: The Professional Drone Pilot's Manual.*

Drone Functionality

Airworthiness

There's a lot more to *airworthiness* than maintenance alone. Human error, especially when dealing with intricate and complex aircraft such as drones is an important issue in itself. If you couple human error with the concepts of airworthiness and maintenance, what you get is a clear need to put systematic safety processes in place.

Most major issues can usually be attributed to insufficient knowledge and a lack of experience in how the UAV should be operated. In order to *Master Flight Techniques* you must constantly update your knowledge base and make a conscious effort to stay ahead of the curve.

Keeping a detailed maintenance log is to become a point of focus in the future as pilots will surely be required to provide them as proof of due diligence and maintaining airworthiness as well as to stay compliant with future regulations. All drone pilots should already complete a routine and properly logged maintenance schedule in order to ensure your aircraft is in top condition. If you do not currently own a UAV Pilot's Flight Time Logbook, I suggest you purchase **Drones: The Professional UAV Pilot's Flight Time and Maintenance Logbook.**

What to Look/Listen out for

Visually check your drone's motors for FOD (Foreign Object Debris) that may have gotten into your drone's motors. Before flights visually check motors in case some of your flight case's foam fitting has gotten in to them. This happened to me once, causing me to waste 20 minutes on site picking polystyrene from my rotor system with a toothpick.

If your drone's motors are making sounds of a higher than normal pitch, this means that for one reason or another your motors are straining, most likely caused by high winds. Fly with caution.

After each flight check the *temperature of your batteries and motors*. If you are changing the battery in order to complete another flight it is all the more important to ensure that the motors are not overheating.

Does your drone have the latest firmware installed? Most drones now require regular *firmware updates* in order to upgrade technical ability, fix bugs and add new features. Always fly with the latest firmware if possible, but I would suggest waiting a week or so after the newest update is released before installing it onto your drone. The week "grace period" will give the developers time to fix any bugs that have arisen in the first few days. Pay close attention to the process that is being used to update the drone, they have to be correct or the update will fail.

Drone Racing

Drone Racing is a high adrenaline competitive sport. Fantastic to watch if your eyes can keep up and even better to participate in, I suggest you all get to work on building your racing drones. In the next few years drone racing is set to sky rocket and professional leagues are already worldwide.

The drones in these races are high performance speed machines and when flown by a skilled pilot the agility of these aircraft is incredible. Being flown in acro or rate mode makes racing drones harder to fly but the improvement to practice ratio means that, within a few months you will be ready to race opponents. An advantage of flying rate mode is that once you have mastered the basics you can begin to develop your own flying style, signature moves and those all-important race winning techniques.

Drone racing is usually flown FPV style or First Person View. By using FPV goggles a pilot's view is as if they are the drone (similar to sitting in the cockpit). This allows for out of sight flying which is necessary for multi-level courses and is the only way to practically compete.

The courses often cover multiple levels that are filled with obstacles to manoeuvre. These 3D race tracks are straight out of science fiction and are flown through at speeds of up to 100mph. Pulling off manoeuvres like 90° turns at these speeds takes considerable skill but for those that can, there is the potential to become superstar. It's still early days for drone racing so it's anyone's game. Within competition races all the drones taking part will have set specifications, but with that being said, drone racing opens up another very interesting area…building your own drone. A topic that will be covered later in the series in ***Drones: Building your own drone.***

When designing a course on which you can practice your racing tactics and techniques, really test your reflexes and your drone's agility, push yourself in a safe environment and it will pay dividends in the long run. A well-crafted course should as a minimum contain the following:

- Slaloms
- Chicane corners
- Multiple levels
- Hairpin turns
- Long or sweeping turns
- Obstacles to avoid or fly through

Generally there are 3 classes of racing drone micro mini and open/free:

Micro

With a size of up to 150mm a micro racing drone is best suited to indoor flight. In the micro drone racing class a drone is limited to 2 batteries and has 4 motors.

Mini

In the extremely popular mini class, drones have up to 5 inch propellers, 4 motors and a size of up to 250mm.

Open/Free

The open or free class of racing drone has fewer restricted in order to allow for greater speed. With a size of up to 300mm and 6 inch propellers these machines hit incredible speeds.

Time to Level-up your Skill-Set

Once you have chosen your racing drone it is time to get out there and join a club. Drone racing for some is all about the speed whilst other pilots favour extreme tricks and gravity defying stunts. The best practice (in my experience) is freestyle practice in a wooded area. It is an amazing experience to fly through a forest FPV, it's a view most people will never get to see and enhances yours skills very quickly. The high adrenaline nature and instant jumps from risk to reward through pulling off spectacular tricks is unbelievably addictive.

Prepare for impact

Crashes happen when you least expect it.

If the worst should ever happen and your drone makes an unscheduled landing, how you react is crucial not only in the recovery of your drone but also in keeping anyone in the surrounding area safe. Below I have set out step by step how to handle the situation:

- Direct everyone in the proximity of the crash site to keep a safe distance.

- Immediately cordon off area.

- If there is any sign of smoke or fire be sure to stay well back.

- Alert the emergency services where necessary.

Commercial drone operators must always follow the Emergency Procedures as per their FRCs (Flight Reference Cards) and set out by the Civil Aviation Authority. This includes gathering evidence and incident reporting. For more information on FRCs and how to author your own, please read the first book in the series *Drones: The Professional Drone Pilot's Manual.*

TIPs:

If you are about to crash immediately turn down the throttle in order to cause as little damage as possible. Should you crash, switch off your remote control straight away (unless otherwise stated in your UAV/drone instructional manual), this should deactivate any motors and stop them overheating after a collision.

Before touching your drone be sure to have your PPE (Personal Protective Equipment) to hand:

- Heat/fire proof gloves
- Fire extinguisher
- Safety goggles
- Safety tape to cordon off crash site
- First aid kit

Take photos of the crash site.

If the drone has crashed into a tree or other high spot, only retrieve it if it is safe to do so.

The first thing you should do, providing the crash site is safe and the rotors have stopped turning is to remove your drone's battery and place it to one side inside of a lipo safe bag.

Disconnect any removable parts from your aircraft and individually inspect them for signs of damage.

Clear any crash debris.

Before you leave the site ensure the crash site is left as it was before the crash, double check that you are taking all crash debris/evidence with you.

At home, go back over the incident and take notes on what caused the crash, human error, drone malfunction, loss of data control link? Once you have detailed notes on what caused the crash make further notes describing how you will stop the same issue arising again or at least how you will next time mitigate the circumstances that caused the previous crash.

An airworthy drone and a mind-set of good airmanship alone will absolutely lower any potential risks; these along with a little planning will assure you're doing everything reasonably possible to be a responsible drone pilot. Have fun be safe and remember; if you haven't fallen off your horse you haven't been riding long enough.

Further Tips and FAQs

Do you need a licence/permit to fly a drone?

In most countries to fly a drone as an enthusiast does not require a licence, qualification or permit/permissions. Qualifications and permissions come into play when a commercial project takes place. In the UK if you are not getting paid for flying your drone, as long as you are compliant with safety regulations, you won't need a "licence" or CAA permissions.

Here are the basic rules:

- Unable to fly 5km near airports.

- Flight limited to below 400ft vertically.

- Flight limited to 500m horizontally.

- Unable to fly within 150m of any congested area.

- Unable to fly within 50metres of any person, vessel, vehicle, or structure not under the control of the pilot in charge.

How high can you fly?

CAA regulations have set a UAV's maximum flight altitude at 400 feet (120 metres). However in special circumstances; applications can be submitted to the CAA to request permission to fly above the ceiling if certain safety precautions are met.

How Far can you fly?

Again CAA regulations have set a radius limit of 500 metres but as the above the limits can be extended by the CAA in special circumstances.

How close to people and buildings can you fly?

Flying UAV's overhead of members of the public without their prior knowledge and agreement is not allowed. Between the UAV and any individual or property not under the pilot's control, a clear radius of 30meters must be maintained during take-off and landing as well as maintaining a 50meter distance during flight.

The 30 meter radius for take-off and landing doesn't apply for people or property under the pilot's control or involved in the shoot, as they will be briefed on the relevant safety precautions.

Inside the UK, if flying within 150 metres of built-up areas or near large crowds Commercial UAV Operators must make an application to the Civil Aviation Authority for clearance.

Will any captured footage be stable?

This depends on your chosen model of UAV but generally speaking, given weather and wind conditions are favourable your drone will be as stable as an industry standard jib. A jib in cinematography is a long arm that operates like a see saw and has a camera fixing on one end; this device allows for greater manoeuvrability than filming by hand alone.

Tips

Accelerate and decelerate gently, sharp movements will shake the gimbal potentially leaving your captured footage shaky.

Always be aware of your drone's orientation, if the drone if facing you, when you push left on the right stick the drone will fly right and vice versa.

Keep your drone batteries in a warm place prior to being used, in cold weather try keeping hand warmers in your drone battery case.

Always carry spare batteries as well as spare connection cables for tablets.

Purchase a travel case for your drone.

If your drone is used for filming always keep a lens pen with you in order to clean the drone's camera lens to avoid raindrops or smudges spoiling your end footage.

Stock up on spare SD cards and batteries.

Replace propellers and their connection points regularly.

Wear a lanyard attached to your remote control to ensure that if dropped it doesn't hit the floor.

Wear lightly tinted sunglasses when flying to prevent glare affecting your ability to fly.

Pay extra attention when completing *firmware updates*.

Join a forum, I have found them to be a wealth of information and filled with knowledgeable individuals who are always willing to quickly provide answers to any questions.

Join a club, drone flying clubs are in every city and are the best place to go to practice your skills in a safe environment alongside other aircraft users.

Finally, always read the instructions.

Afterword

You made it!

Congratulations on reaching the end of the *Drones: Mastering Flight Techniques*, the second book in the series *Drones*. I truly hope the information in this book helps you in your journey towards becoming the best drone pilot you can be. Please look out for the soon to be released, next book in the series, *Drones: Building Your Own Drone*.

If you have any comments, did you enjoy the book? Was the information useful? Did I miss anything? Please let me know via Amazon review.

Best Wishes

Brian Halliday

Printed in Great Britain
by Amazon